HOW TO … CRYSTALS

HONEY ANDERSEN

MONDO

This edition first published in the United States in 1997 by
MONDO Publishing
By arrangement with MULTIMEDIA INTERNATIONAL (UK) LTD

Acknowledgments
Grateful acknowledgment is made to the Museum of Victoria, Australia, for the following photographs, reproduced with permission through Multimedia International (UK) Ltd: saleeite—Lake Boga, Victoria: covers; cerussite—Broken Hill, New South Wales: pp. 3, 4, 5; actinolite—Dookie, Victoria: pp. 6, 7 (top); natrolite with analcime and dolomite—Cape Shanck (near Flinders), Victoria: pp. 8, 9 (top), 10, 11 (top), 12, 13 (top), 14, 15 (top right and bottom right), 16, 17 (top); siderite on blue clay with phillipsite—Narre Warren, Victoria: pp. 20, 21 (top), 22, 23 (top).

Evan Clark: photographs pp. 7 (bottom), 9 (bottom), 11 (bottom), 15 (top left and bottom left), 17 (middle and bottom).

Lynn Twelftree: diagrams pp. 11, 19, 20, 21, 23.

For information contact:
MONDO Publishing
980 Avenue of the Americas
New York, NY 10018
Visit our website at www.mondopub.com

Printed in China
First Mondo printing, October 1996
11 12 9 8

ISBN 1-57255-196-8

Originally published in Australia in 1987 by Horwitz Publications Pty Ltd
Original development by Robert Andersen & Associates and Snowball Educational

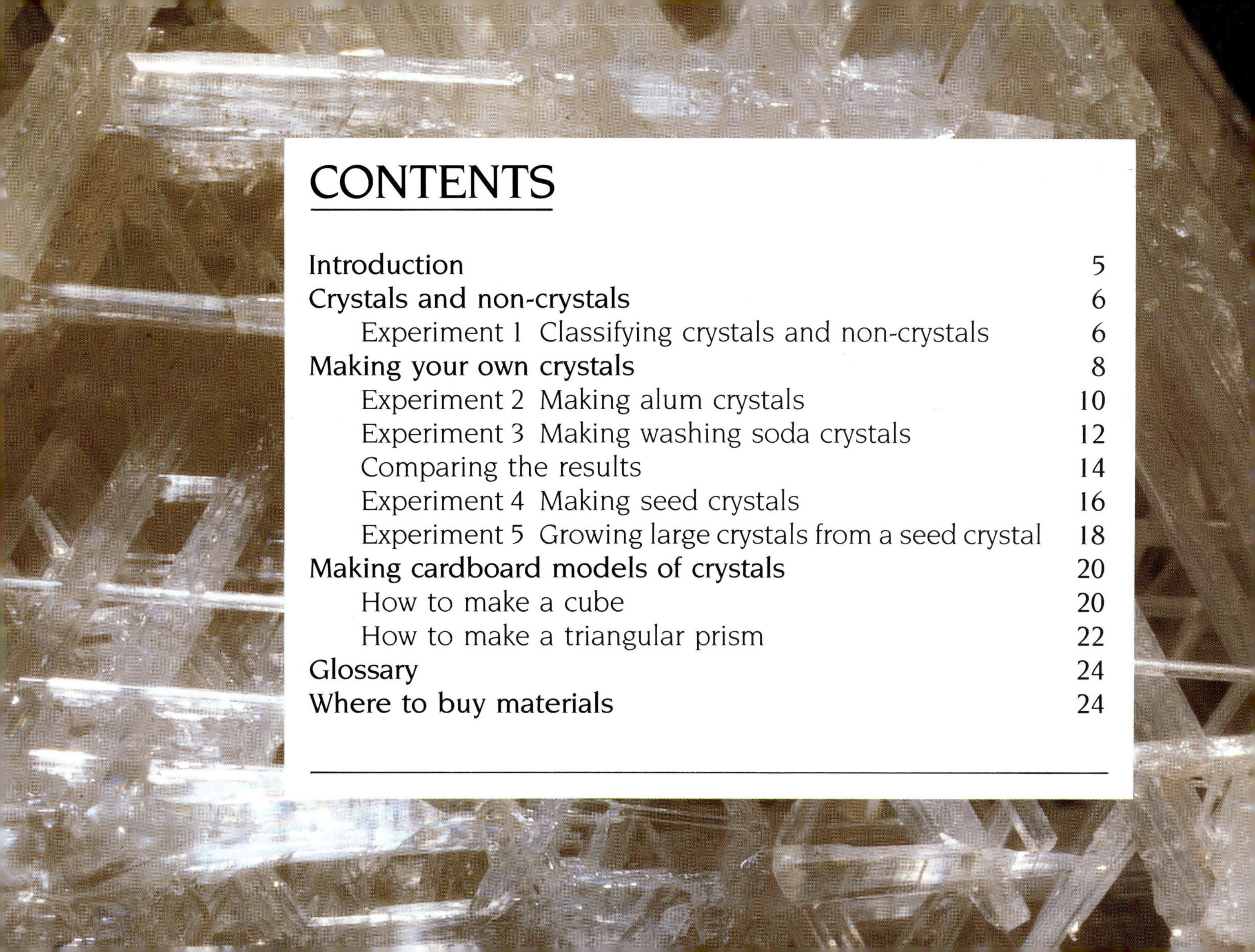

CONTENTS

INTRODUCTION

Some materials that are solid are called crystals. Crystals can be different sizes and shapes but most have flat faces (or surfaces) and straight edges.

In this book you will find several experiments for making crystals.

WARNING **Be careful when handling the materials.**

- Do not eat any of the materials or drink the solutions.
- Do not touch your eyes or mouth during the experiments.
- Wash your hands as soon as you have finished each experiment.
- Be careful when using hot water.
- Be sure to have an adult help you.

CRYSTALS AND NON-CRYSTALS

EXPERIMENT · 1 *Classifying crystals and non-crystals*

You will need:

Equipment

1 magnifying glass
1 pencil
1 piece of paper

Materials

1 teaspoon of sugar
1 teaspoon of unsweetened cocoa
1 teaspoon of salt
1 teaspoon of sand
1 teaspoon of tea leaves

Method

1 On your paper draw two columns with the headings Crystals and Non-crystals. Sort the materials into what you think are crystals and non-crystals. Write their names under the headings.

2 Now, using the magnifying glass, check each material in turn to see whether or not the particles look like crystals.

3 Check your list. Were you right?

MAKING YOUR OWN CRYSTALS

In the following experiments you are not told the exact amount of each material to use. This is because the amount that will dissolve will vary depending on how hot the water is and how quickly you stir it in.

Always use water that is very hot, but not boiling. Add the material that is to be dissolved, one teaspoonful at a time, and stir it until it has completely dissolved before you add the next one. Keep a record of how many spoonfuls you use.

PICNIC PACK
SAXA table SALT
125 g NET
Sigma
ALUM
375 g net
Sugar Babies
SPECIAL OFFER
CSR
WHITE SUGAR
SOOTHING
Lectric Sod
WASHING & BATH CRYSTALS
Relief for tired feet -weary, aching limbs.
750g NET
REFRESHES AND
salt
alum
sugar
washing soda

EXPERIMENT · 2 *Making alum crystals*

You will need:

Equipment

1 jam jar
1 metal teaspoon
1 pencil
1 paper clip
1 piece of thread

Materials

alum
$3\frac{1}{2}$ oz. (100 ml) of hot water

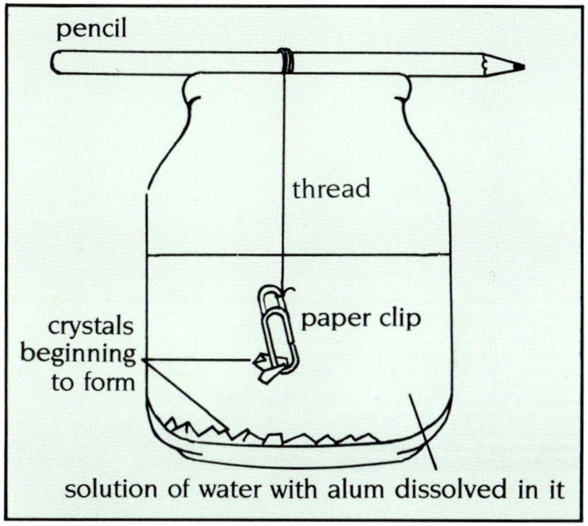

Method

1 Tie one end of the thread to the paper clip and tie the other end to the pencil.

2 Pour the hot water into the jar. Stir alum into the water, a teaspoonful at a time, until no more alum will dissolve in the water.

3 Balance the pencil over the jar so that the paper clip hangs down inside the jar and is in the solution.

4 Put the jar where it will not be disturbed and check it every day. Very soon crystals will form on the paper clip and the bottom of the jar.

EXPERIMENT·3 *Making washing soda crystals*

You will need:

Equipment

1 jam jar
1 metal teaspoon
1 pencil
1 paper clip
1 piece of thread

Materials

washing soda
$3\frac{1}{2}$ oz. (100 ml) of hot water

Method

1. Tie one end of the thread to the paper clip and tie the other end to the pencil.
2. Put the hot water into the jar and then stir in the washing soda, a little at a time, until no more will dissolve.
3. Balance the pencil over the jar so that the paper clip hangs down inside the jar and is in the solution.
4. Put the jar where it will not be disturbed and check it every day. Watch the crystals form.

Comparing the results

1 Compare the size of a washing soda crystal to the size of an alum crystal.
- Which crystal did you think would be bigger?
- Were you right?
- Where were there more crystals – on the thread or on the side of the jar?

2 Now try the same experiment using salt or sugar. Try to guess how big the crystals will be and how long they will take to form.

3 Draw up a chart like this:

Crystals	Size of crystals	How long they took to grow
Alum		
Salt/Sugar		
Washing soda		

4 Compare the results of your experiments.

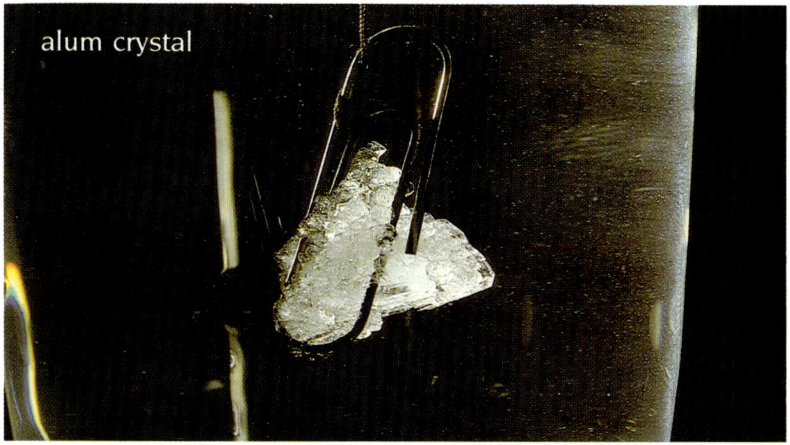

alum crystal

washing soda crystal

EXPERIMENT · 4 *Making seed crystals*

You can grow bigger crystals than the ones you have already made by first making "seed" crystals.

You will need:

Equipment

1 saucer
1 jam jar
1 metal teaspoon

Materials

alum
2 oz. (60 ml) of hot water

Method

1. Measure the water into the jar and stir alum into it, little by little, until the alum no longer dissolves.
2. Tip half the liquid onto the saucer and put it where it will not be disturbed for 1-2 days.
3. The liquid in the saucer will crystallize.
4. Choose four or five of the largest, best-shaped crystals to be the "seed" crystals.

EXPERIMENT·5 *Growing large crystals from a seed crystal*

You will need:

Equipment

1 cup of hot water
1 jam jar with a lid
1 piece of thread
1 pencil
1 metal tablespoon
1 coffee filter paper

Materials

1 seed crystal
4 tablespoons of alum

Method

1 Put the water into the jar and stir the alum into it. Cap the jar and let it stand for 24 hours, shaking it every now and then until as much alum as possible has dissolved.

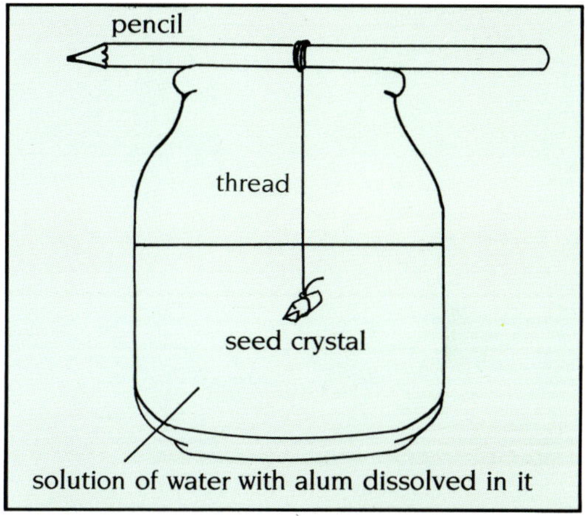

2. Pour the alum solution through the filter paper to remove the extra grains. Then return the solution to the jar.
3. Tie one end of the thread around the seed crystal. Tie the other end of the thread around a pencil.
4. Balance the pencil so that the "seed" is in the solution.
5. Put the jar in a cool place, where it will not be disturbed. Watch the crystals grow.

Comparing the results

Measure the time it takes for a large crystal to grow and compare this with the time it took to grow the smaller crystals.

MAKING CARDBOARD MODELS OF CRYSTALS

How to make a cube

You will need:

1 pencil
1 ruler
1 sheet of thin cardboard
tape
scissors

What to do

1 Draw a rectangle that is 8 in. (20 cm) long and 2 in. (5 cm) wide.

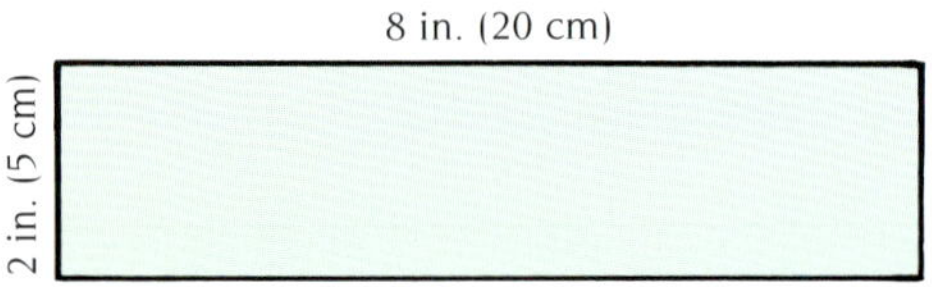

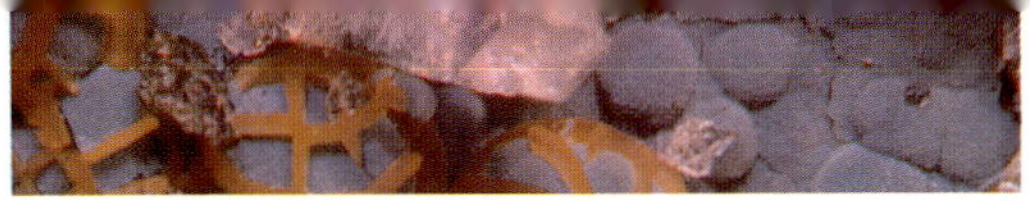

2 Draw lines across the rectangle every 2 in. (5cm), so that it is marked into four equal squares.

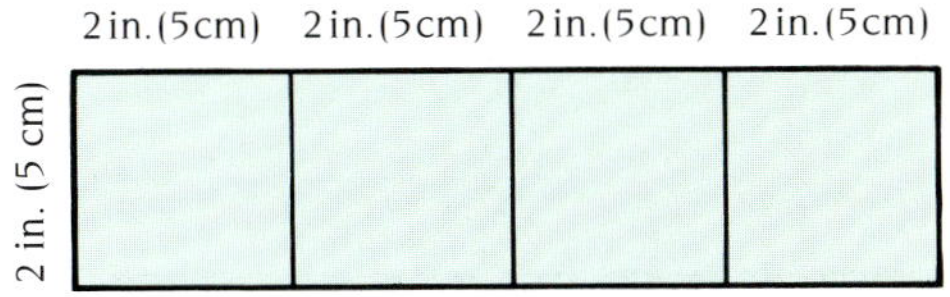

3 Draw a 2 in. (5cm), square at the top and bottom of the second square of the rectangle.

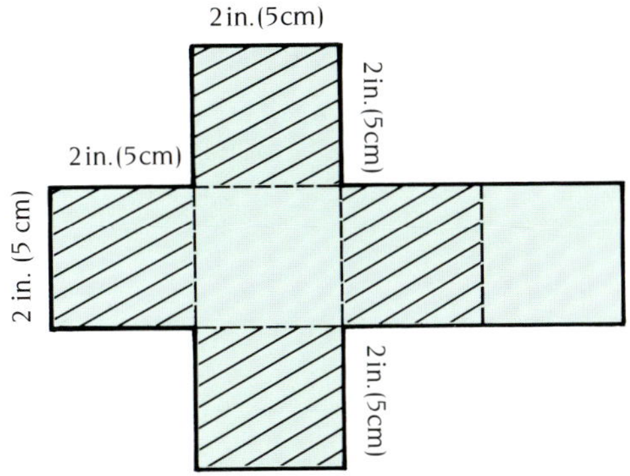

4 Cut around the edge of the pattern.
5 Make a crease along each of the inside lines.
6 Fold along the inside lines.
7 Use tape to join the edges together.

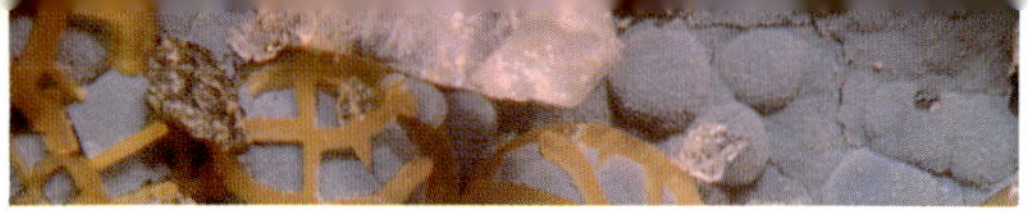

How to make a triangular prism

You will need:

1 pencil
1 ruler
1 sheet of thin cardboard
tape
scissors

What to do

1. Draw a triangle that is 6 in. (15 cm) on each side.
2. Mark it into four smaller equal-sized triangles, as the picture shows.
3. Cut around the edge of the big triangle.
4. Make a crease along each of the inside lines.
5. Fold along the inside lines.
6. Use tape to join the edges together.

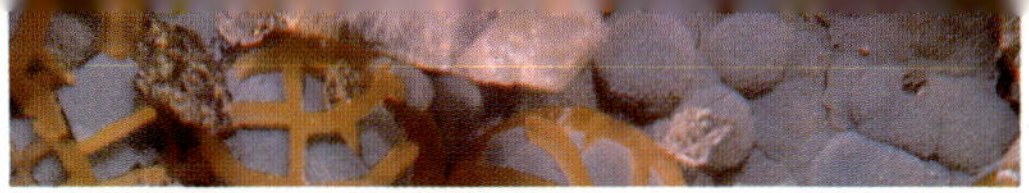

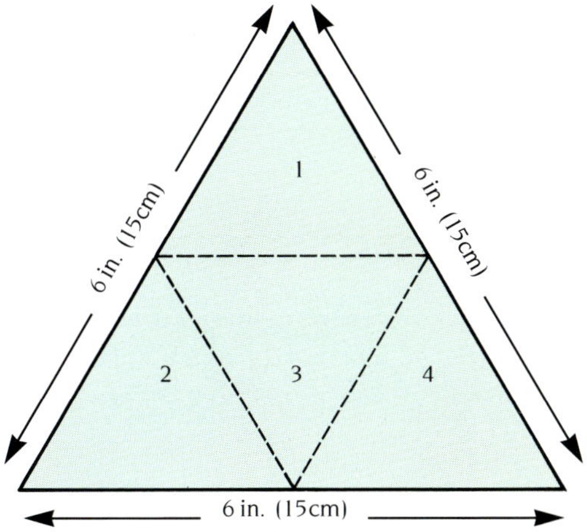

Can you find some crystals that are the same shape as these models?

GLOSSARY

crystallize	To form crystals. When a solution has crystallized it is no longer a liquid. It is a solid.
cube	A solid square shape. A cube has six faces and each face is a square.
dissolve	To combine with water. A solid **dissolves** when it is put into water and combines with the water to become liquid.
particle	The smallest piece. For example, a grain of sand is a particle.
seed crystal	A small crystal used to make a larger one.
solution	Water that has some solid matter dissolved in it.
triangular prism	A solid triangle shape. It has four faces and each is a triangle.

Where to buy materials

The materials used to make crystals in this book are available from pharmacies. Most supermarkets also sell washing soda.

Pharmacists may also suggest other materials that would make different shaped crystals and are safe to use.